W9-CZR-473

TEAM EARTH

DECOMPOSERS AND SCAVENGERS

NATURE'S RECYCLERS

BY EMMA HUDDLESTON

CONTENT CONSULTANT
Thomas Volk, PhD
Professor of Biology
University of Wisconsin-La Crosse

Cover image: Mushrooms commonly act as
decomposers, returning nutrients to the soil.

Core Library

An Imprint of Abdo Publishing
abdobooks.com

abdobooks.com

Published by Abdo Publishing, a division of ABDO, PO Box 398166,
Minneapolis, Minnesota 55439. Copyright © 2020 by Abdo Consulting
Group, Inc. International copyrights reserved in all countries. No part of this
book may be reproduced in any form without written permission from the
publisher. Core Library™ is a trademark and logo of Abdo Publishing.

Printed in the United States of America, North Mankato, Minnesota
102019
012020

Cover Photo: Jaroslav Machacek/Shutterstock Images
Interior Photos: Jaroslav Machacek/Shutterstock Images, 1; A. Michael Brown/Shutterstock
Images, 4–5; Jacob Eukman/iStockphoto, 7; Kazakova Maryia/Shutterstock Images,
11 (background and tree); Shutterstock Images, 11 (horse), 22–23, 26, 39 (apples), 39 (dirt);
iStockphoto, 12, 29, 40, 45; Marie Lemerle/Shutterstock Images, 14–15; David Litman/Shutterstock
Images, 16, 43; Karel Bartik/Shutterstock Images, 18; Sergey Uryadnikov/Shutterstock Images,
20; Andrea Geiss/Shutterstock Images, 27; Bogdan Ionescu/Shutterstock Images, 30–31; Eye of
Science/Science Source, 34; Dmytro Ostapenko/Shutterstock Images, 36

Editor: Marie Pearson
Series Designer: Megan Ellis

Library of Congress Control Number: 2019942106

Publisher's Cataloging-in-Publication Data

Names: Huddleston, Emma, author.
Title: Decomposers and scavengers: Nature's Recyclers/ by Emma Huddleston
Other Title: nature's recyclers
Description: Minneapolis, Minnesota : Abdo Publishing, 2020 | Series: Team earth | Includes
        online resources and index.
Identifiers: ISBN 9781532190988 (lib. bdg.) | ISBN 9781644943250 (pbk.) | ISBN
        9781532176838 (ebook)
Subjects: LCSH: Decomposition (Biology)--Juvenile literature. | Biodegradation--Juvenile
        literature. | Food webs (Ecology)--Juvenile literature. | Ecology--Juvenile literature.
        | Organic wastes--Juvenile literature.
Classification: DDC 577.16--dc23

CONTENTS

CHAPTER
ONE

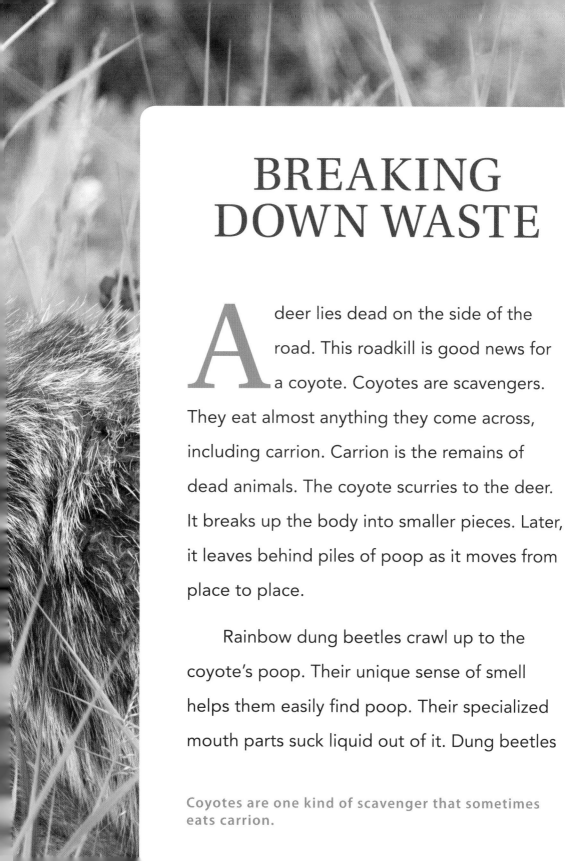

BREAKING DOWN WASTE

A deer lies dead on the side of the road. This roadkill is good news for a coyote. Coyotes are scavengers. They eat almost anything they come across, including carrion. Carrion is the remains of dead animals. The coyote scurries to the deer. It breaks up the body into smaller pieces. Later, it leaves behind piles of poop as it moves from place to place.

Rainbow dung beetles crawl up to the coyote's poop. Their unique sense of smell helps them easily find poop. Their specialized mouth parts suck liquid out of it. Dung beetles

Coyotes are one kind of scavenger that sometimes eats carrion.

COMPOSTING AT HOME

Composting is a way to recycle food and yard waste instead of adding it to piles of trash. People with backyards can make a compost pile there. People can also compost in a plastic bin after putting soil in it. Successful composting has three parts: browns, greens, and water. Browns include waste such as dead leaves and sticks. Greens include waste such as grass and food scraps from fruits and vegetables. Water keeps the soil wet and speeds up decomposition. Bacteria in the soil decompose the organic waste. The bacteria turn the waste into nutrients. The compost can be added to soil to help plants grow.

eat poop. They take in leftover nutrients that larger animals don't digest. Nutrients are substances that help living things grow and give them energy. Some kinds of these beetles live in the waste. Others roll it into a ball. Females lay one egg inside each ball and bury it. The young beetle is surrounded by food as it grows. Dung beetle behavior may seem gross, but it is important. It exposes

Dung beetles live on six of the seven continents. They do not live in Antarctica.

SAVING ANCIENT ARTIFACTS

Bob Blanchette is a mycologist. He studies fungi. Fungi are important decomposers, but their ability to break down wood has damaged ancient artifacts. In Egypt, fungi have ruined ancient coffins. They grow in the wood. They make millions of tiny holes. If the coffin is touched or moved, it turns into dust. Blanchette is working with historians to save the coffins. By studying the fungi first, they can figure out how to stop the wood from becoming dust. This will keep ancient artifacts from being destroyed.

the poop to bacteria and fungi in the soil.

Throughout this process, bacteria and fungi have been at work. Bacteria and fungi are decomposers. Decomposers break down waste into nutrients. They are different from scavengers, who also help break down waste but don't turn it into nutrients. Grasses growing nearby benefit from the bacteria's activity. They use nutrients in the soil as food. A deer and her fawn graze on these grasses. The deer get nutrients.

DECOMPOSERS IN THE FOOD WEB

Decomposition is the process of breaking down waste. Waste is dead organic matter. This matter is most often in the form of leaf litter, dead wood, poop, and dead animal bodies. Decomposition can happen quickly, or it can happen slowly over time. The speed depends on the amount of waste, the number of decomposers at work, and the environment. Waste is food for many living things.

The main difference between scavengers and decomposers is what happens to the waste after they eat it. Scavengers quickly break large amounts of waste, such as dead bodies, into smaller pieces. Vultures, some insects, and worms are scavengers. Scavengers leave behind more waste, though it is more broken down than its original form. Decomposers turn that waste into nutrients. The nutrients return to the soil, water, and air.

The food web is a cycle of nutrients and energy in an environment. Living things fall into three roles in a food web. These roles are producers, consumers, and decomposers. Decomposers are a major part of the food web because they recycle organic matter. They usually produce the nutrients carbon, nitrogen, and water. Plants use these simple nutrients to grow. Plants are producers. They get energy from sunlight. Consumers are the third major part of the web. Most consumers are animals. They get their nutrients and energy from eating plants or other animals. Scavengers are a type of consumer that eats dead plants or animals.

HEALTHY ENVIRONMENT

Dung beetles eat, collect, and bury waste. They quietly remove waste from land before it becomes a problem. For example, farmland in Australia in the 1960s suffered from too much cow manure. The poop dried in the sun and killed the grass below it. Worms and flies infested the poop. They spread diseases to the cows. Then dung beetles came to the rescue. Scientists brought the beetles to the area. They helped bury the waste and create balance in the environment.

NUTRIENT CYCLE

Decomposers play an important role in recycling nutrients. They break down organic matter. They put nutrients back into the soil. Plants use these nutrients to grow. Decomposers help the whole environment survive because plants are food for many animals. How does this diagram help you understand decomposers' role in the food cycle?

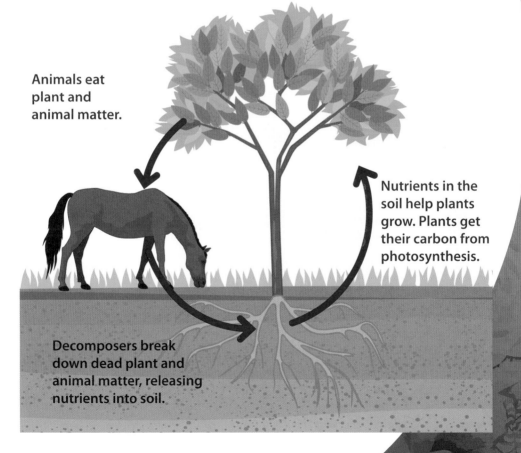

Animals eat plant and animal matter.

Nutrients in the soil help plants grow. Plants get their carbon from photosynthesis.

Decomposers break down dead plant and animal matter, releasing nutrients into soil.

Many species of decomposers are tiny. Some can only be seen under a microscope. But together, the power of many small species adds up. Decomposers are a necessary part of a healthy environment. Without them, ecosystems suffer. Dead animal bodies would never go away. Forests would be full of fallen trees and leaves instead of healthy plants. Farmland would be covered in animal poop instead of crops. All people around the world benefit from the work of nature's recyclers.

FURTHER EVIDENCE

Chapter One talks about what decomposers do. Identify the main point and some key supporting evidence. Then look at this website. Find a quote that supports the chapter's main point. Does the quote support a piece of evidence already in the chapter? Or does it add a new piece of information?

WHAT IS A DECOMPOSER?

abdocorelibrary.com/decomposers-and-scavengers

Fungi help decompose dead trees.

BIG SCAVENGERS

Scavengers feast on dead plants and animals. They are not decomposers, but they are sometimes involved in early stages of the decomposition process. They break large pieces of organic matter into smaller pieces. When scavengers finish eating, they leave smaller pieces of waste and poop behind. At that point, everything is up to decomposers. Scavengers come in many shapes and sizes. Birds, mammals, and sea creatures can all be scavengers.

Sometimes multiple scavengers feed on the same carcass.

Turkey vultures are a common species of vulture in North America.

VULTURES

Vultures are scavengers. They eat carrion. Vultures' adaptations make them effective scavengers. They have a strong immune system to protect their bodies from getting sick after eating rotting meat. Their heads and necks are bald. Carrion flesh, which may carry diseases, cannot stick as easily to bare skin as it can to feathers.

Vultures' large wings let them travel far without flapping much. From the air, vultures can see lots of land below. They can spot a dead animal 20 miles (30 km) away.

Most vultures don't know when they will get their next meal. For this reason, they stuff themselves when they eat. Sometimes they are so full they can't fly. A throat pouch lets them store even more food for later.

TONS OF DECOMPOSING PIGS

In 2017, a team of scientists studied how 3 tons (2.7 metric tons) of dead pigs decomposed in a forest in Mississippi. They placed the dead pigs there to study unexpected, natural mass-death events. They found the environment changed a lot. An unusually high number of coyotes, vultures, and armadillos came to eat the bodies. The leaf litter was soaked in a slimy substance. The soil was full of substances and insects not normally found there. For all these reasons, one scientist thought the location would never be the same as it was before.

The lammergeier vulture lives in the mountainous regions of Africa and Eurasia.

Different kinds of vultures scavenge in different ways. Lammergeiers are a type of vulture that eats bones. They drop bones from high in the sky. The bones break on rocks below. Then the birds eat the soft tissue inside.

EATING ANYTHING

Great white sharks are powerful hunters. When food sources are low, they also scavenge. They eat dead whales, fish, and sea lions. Many scavengers have a flexible diet. They don't rely fully on carrion like vultures do. Hyenas, coyotes, and polar bears can hunt or scavenge.

Most scavengers eat whatever they can find. Racoons and seagulls will eat food out of garbage cans. Eagles and crows often hunt small animals or fish. In the winter, it can be difficult to find food. Then eagles and crows will eat carrion too.

HYENAS

Hyenas are not strictly scavengers. A group of hyenas, called a clan, often hunts for its food. They work together to take down prey. A lone hyena often eats animals that died naturally or were killed by another animal. Without a clan to help, it is much more dangerous trying to take down an antelope, whose hooves could injure a hyena.

Many scavengers, including great white sharks, are also skilled hunters.

Hyenas, coyotes, and wolves will also eat carrion when prey is scarce, or hard to find.

Crabs and lobsters crawl on the bottom of the sea floor. This location limits their food sources. They often eat dead fish and shrimp because bodies sink right after death. Scavengers' ability to eat dead or living food helps them survive. They can live in places with limited food sources. They can also adapt easily to new environments. They carry on their important work around the world.

STRAIGHT TO THE
SOURCE

Christopher O'Bryan and James Watson study the environment and conservation. They wrote an article about how important predators and scavengers are:

> *Animals host over 60 percent of known human diseases, and predators and scavengers help prevent their spread by consuming the animal hosts of these ailments. . . .*
>
> *Insect-eating frogs may play a global role in reducing dengue fever by preying on mosquito eggs. Similarly, vultures and other scavengers may substantially reduce disease spread by rapidly consuming carcasses that would otherwise benefit stray dogs, rodents and other disease carriers.*

> Source: Christopher O'Bryan and James Watson. "We Should Embrace Scavengers and Predators." *Scientific American*. Scientific American, March 9, 2018. Web. Accessed August 2, 2019.

Back It Up

The author of this passage is using evidence to support a point. Write a paragraph describing the point the author is making. Then write down two or three pieces of evidence the author uses to make the point.

SMALL SCAVENGERS

Not all scavengers are as large as a vulture, shark, or crab. Scavengers can also be very tiny. Worms and insects are very important scavengers.

KING OF SMALL SCAVENGERS

Earthworms are excellent scavengers. More than 1,800 species exist. They can live almost anywhere. They thrive in humid environments. They spend their whole lives helping their environment by eating soil and the organic matter in it. They only use 27 percent of the nitrogen in their food. The rest goes out of their body in the form of a cast. Casts are

Earthworms eat organic matter in the soil.

RADIATION AND DECOMPOSITION

Timothy Mousseau is a biologist studying decomposition at the Chernobyl nuclear power plant in what is now Ukraine. In 1986, part of the Chernobyl plant exploded. This let a massive amount of radioactive material into the air. Radioactive material is dangerous because it weakens cells. Weak cells break down easily. This can lead to growth deformities, cancer, or death. Mousseau and other scientists scattered mesh bags of leaves and other natural litter around Chernobyl. This let them see how areas with different levels of radiation did or didn't decompose. They found radiation slows down decomposition. Less nutrients return to the ground, slowing plant growth.

earthworm waste. Casts make soil healthy because they are rich in nutrients.

An earthworm's movements also loosen the soil. It wiggles its soft body and burrows into the ground. Air can easily pass through loose soil, which helps plants grow. The tunnels worms dig create space for rainwater to soak deep into the soil.

Worms can be grouped based on how deep they live in

the soil layers. Some worms live on top of the soil and in leaf litter piles. They eat leaves and fallen organic matter. Many light pink and white worms live in the humus. Humus is the dark top layer of soil. Since these worms get little sunlight, their skin is light-colored. It can be light pink, gray, or white. They eat organic matter below ground, such as dead plant roots. Many worms that are active at night live several feet down into the soil. They pull leaf litter into their deep burrows to feed. Some night crawlers can travel up to 62 feet (19 m) on the soil surface. They look for food before they dive back down.

BITING, FLYING, AND CRAWLING

Many insect species are decomposers. Springtails break down organic matter in soil all over the world. They are tiny shrimp-like insects. More than 6,000 species of springtails exist. They are one of the most numerous insects on Earth. Scientists think that in some regions, springtails decompose up to 20 percent of the fall leaf litter.

Springtails are found on all seven continents.

Maggots eat dead animal flesh. Maggots are fly larvae. Some flies also eat dead animals. Large numbers of them swarm to dead bodies. Blowflies eat rotting plant matter and garbage. Sometimes they also bite open wounds on living animals. Cockroaches eat dead animals, plants, paper, and other rotting matter.

Maggots help break down animal remains.

Other small scavengers include millipedes. Millipedes crawl on the top layer of soil. Those that live in tropical areas grow larger than 8 inches (20 cm). Millipedes feed on plant matter and organic litter. They spread their waste on the soil surface as they move.

BEETLES AND TERMITES

Many beetles live in dead and dying trees. They can help with early stages of wood decomposition. Other beetles eat flesh. However, the most famous beetle scavengers are dung beetles. They eat, roll away, and

BEETLES HELP SOLVE CRIME

Dermestid beetles are flesh-eating beetles. They can eat a dead body down to its bones. This process is called skeletonization. Dermestids' eating habits are valuable for multiple reasons. Museum displays and scientific research often need clean bones. Sometimes crime investigations need to look if marks were left on a skeleton. Dermestids are a natural way to clean a body until only the bones are left. The beetles don't damage the bones. Without dermestids, scientists would have to use chemical substances to remove the flesh. Chemicals can damage the bones.

bury poop. Their waste removal activities are extremely important.

Termites live in tropical areas, deserts, and forests. They are one of the most important parts of their environments because they can eat and break down wood waste. Wood is a very difficult type of waste to get rid of. The more termites there are in an area, the faster wood gets broken down. Wood is not easy to digest.

Termites rely on bacteria in their guts to break down the wood they eat. Bacteria are made of single cells.

Some dung beetles can roll a ball 50 times heavier than the beetle itself.

After termites do their job, fungi finish decomposing the wood.

Termites, ants, and worms are the three major Earth-moving insects. They are small, but there are a lot of them. They are key to balancing the cycle of nutrients and energy in their environments.

EXPLORE ONLINE

Chapter Three talks about earthworms. Visit the website below. Compare and contrast the information there with information from this chapter. What new information did you learn?

NATIONAL GEOGRAPHIC KIDS: EARTHWORMS

abdocorelibrary.com/decomposers-and-scavengers

BACTERIA AND FUNGI

Bacteria and fungi, including mold, are key decomposers. They can break down waste at any stage of the decomposition process. They work in nearly all environments. Some can survive harsh conditions, such as extreme heat or cold. Some can live in water or deserts. This range of habitats makes them important for recycling nutrients all over the world.

When decomposers break down leaf litter, they take part in the carbon cycle. The carbon cycle is the movement of carbon throughout the air, water, and ground. Decomposers

Mushrooms are one recognizable part of fungi. But the majority of the fungus a mushroom grows from is underground or in wood.

recycle carbon and move it from one place to another. Humus is long-term carbon storage for many environments. Carbon can be stored there for hundreds or thousands of years.

The amount of carbon in an environment changes daily. One of the nutrients decomposers create from leaf litter is carbon dioxide. Carbon dioxide is a combination of carbon and oxygen. Plants use carbon dioxide in photosynthesis. They pull it from the air. Then they release the oxygen. Decomposers, humans, and other animals use that oxygen. The cycle continues.

TINY BACTERIA

Bacteria are too small to see with the naked eye. They take in waste as nutrients. They turn the nutrients into energy. Then they use that energy to split in half and take in more nutrients. The process continues. The bacterial population grows rapidly.

Thousands of kinds of bacteria live in soil. They are part of the nitrogen cycle. Some bacteria take nitrogen from the air and turn it into nutrients plants can use. Bacterial activity helps many living things.

Some bacteria live inside living bodies. Some live in termites' guts. Moose, sheep, and deer need other bacteria in their stomachs to help digest tough plant tissue. Bacteria also live in all animals, including humans. Scientists estimate that human cells make up 43 percent of cells in the body. The rest are bacterial cells. Most of these bacteria live in digestive organs. They help break down food so humans can get nutrients from it.

BACTERIA ON THE *TITANIC*

In 2010, scientists found a new kind of bacteria on the *Titanic*. The *Titanic* was a cruise ship that sank in 1912. Today, it lies on the bottom of the Atlantic Ocean. The metal of the ship has rusted over time. Rust is a reddish-brown substance caused by the metal's exposure to oxygen. Bacteria on the ship are eating bits of organic matter in the rust. Eventually, the bacteria could eat the entire ship.

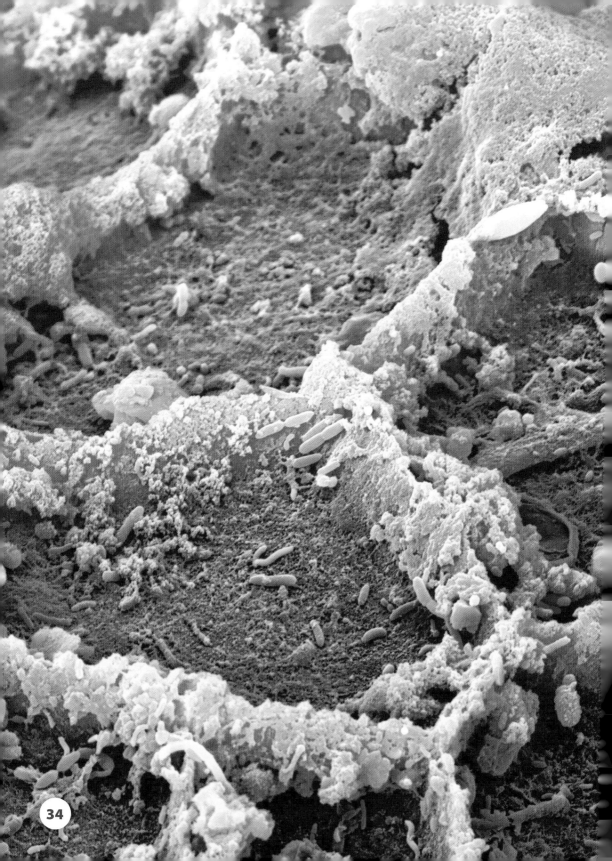

Some kinds of bacteria in the body are helpful. Other kinds are dangerous. If bad bacteria get into a person's body, they can cause diseases.

FUNGI

Scientists estimate there could be 1.5 million types of fungi. Different kinds include mushrooms and mold. Fungi reproduce through spores. These microscopic cells spread through the air and water, as well as on insects and living things. Fungi grow from spores, similar to how plants sprout from seeds.

Plants use sunlight and water to make their own food. This process is called photosynthesis. Fungi are different from plants. They cannot make their own food. They get their energy and nutrients from whatever they grow on. Fungi grow in many substances, including soil, plants, animal bodies, and pieces of food.

Like bacteria, fungi release nutrients for plants to grow. The body of a fungus is called the mycelium.

A magnified image shows bacteria decomposing a leaf.

A block of soil reveals a fungus's mycelium, which looks like white threads.

Its dense threads spread out to take in nutrients and grow. It is the part that decomposes plant tissue and waste. First, the mycelium releases enzymes. The enzymes break down organic matter around the fungus. Then, the mycelium takes in the nutrients. It grows larger. Mushrooms eventually grow from the mycelium. They release spores. Animals that eat mushrooms spread the spores. Mushrooms only appear at certain times of the year. They grow from mycelium similar to how flowers bloom from plants and trees.

Many fungi team up with other species to speed up the cycle of nutrients. One group of fungi lives in plant roots. The fungi and plants help each other. The fungi take in nitrogen and water from the soil. They release the nutrients to the plants' roots. In turn, the fungi get sugars from the plant. These sugars help the fungi grow. Thousands of species of this type of fungi exist.

MUSHROOMS CLEAN UP POLLUTION

In 2015, a study found that a Pueblo community garden in New Mexico had harmful chemicals. Some were chemical waste from industry. Others were fuel for vehicles. The chemicals were spreading to the food growing in the garden. People were getting sick. Tewa Women United, a group of Native women, looked for a solution. The group started growing oyster mushrooms in the gardens. The fungi can filter harmful chemicals from the soil. The group hopes that this will make the soil safe again.

Fungi are the only decomposers that can break down lignin. Lignin is what makes wood so rigid and strong. Bracket fungi decompose wood. They often grow on old trees and fallen logs. They thrive in damp forests. Bracket fungi are also known as shelf fungi because of the shape they grow in. Their flat and wide outgrowth looks like a shelf.

MOLD

Mold is a fungi. Many types of mold grow on food. The spores spread to bread, cheese, meat, and fruit. Mold decomposes food the same way other fungi do. It breaks down food outside the cell. It grows as it absorbs the nutrients. Mold can be dark, white, or green. Some grows into fuzzy patches. Mold can take over food in as little as 12 hours. Other times, it takes weeks to spread. Bacteria and fungi play an important role in the environment. They keep the world clean.

Scavengers and decomposers are all around. Without them, the food web would be incomplete.

STAGES
OF DECOMPOSITION

Decomposers may be tiny, but their work in decomposition can be easy to see. This diagram shows an apple in various stages of decomposition. How does seeing these stages help you understand the cycle of nutrients on Earth?

1.

2.

3.

4.

5.

Mold

6.

Fungi are extremely important in the nutrient cycle.

Eventually, waste would pile up. It would cover Earth and cause harm to plants, animals, and people. Decomposers live in soil, in bodies, and in all types of environments around the world. They recycle nutrients such as carbon and nitrogen. Their activity helps plants grow. Those plants become a food source for most living things. They make life on Earth possible.

STRAIGHT TO THE
SOURCE

Researcher Jason Borchert wrote an article explaining how decomposers are important for the life cycle in freshwater environments:

> *Bacteria are tiny single-celled organisms that can exist in very large numbers in the soil, and to a lesser amount in the water, of freshwater systems. Bacteria are one of the main types of organisms responsible for breaking down dead matter in freshwater systems. . . .*
>
> *Fungi also take part in breaking down dead matter. You can find various types of fungi such as water molds, mildews, and yeast in freshwater systems. And despite fungus's appearance and people's initial beliefs about fungi, fungi are now actually thought to be more closely related to animals than plants.*
>
> Source: Jason Borchert. "Producers and Decomposers of Freshwater." *Ask a Biologist*. Arizona State University, April 5, 2015. Web. Accessed August 2, 2019.

What's the Big Idea?
Read the primary source text carefully. Determine the main idea and the details that support it. Then name two or three of the supporting details.

FAST FACTS

- Decomposers break down waste in their environments. They are one of the three main parts of the food web.

- Scavengers are different from decomposers. Sometimes they are involved in early stages of decomposition. They eat carrion and break down large amounts of waste into smaller pieces.

- Vultures, crows, opossums, and crabs are some scavenger species. Many scavengers eat anything they can find when food sources are low.

- Earthworms and termites are important scavengers. They spend most of their lives eating soil and wood. Bacteria in their guts return nutrients to the soil.

- Bacteria and fungi are key decomposers that live almost everywhere around the world. They can decompose organic matter at any stage.

- Bacteria can live in harsh environments. Some even live inside insect, animal, and human bodies to help digest food.

- Fungi are different from plants because they rely on whatever they grow on for food. Fungi can spread to soil, food, plants, and animal bodies.

- Fungi are the only decomposers that can break down lignin in wood. One group of fungi partners with plant roots to help both the plant and the fungus get more nutrients.

- The actions of insects, bacteria, and fungi may be small, but they make a big difference. Without decomposers, important nutrients would not be moved back to the soil for new life to use.

STOP AND
THINK

Take a Stand

Chapter One explains that decomposers are one of the three main roles in the food web. Producers and consumers are the other two kinds. Do you think one of these three roles is more important than the others? Or do you think they are all equally important? Explain your answer.

Why Do I Care?

Maybe you think decomposition is gross. That doesn't mean you can't think about why it is important. Choose one type of decomposer. How does it recycle waste in its environment? How would the world be different without it?

Another View

Chapter Four talks about the role of bacteria and fungi in decomposition. As you know, every source is different. Ask a librarian or another adult to help you find another source about this process. Write a short essay comparing and contrasting the new source's point of view with that of this book's author. What is the point of view of each author? How are they similar and why? How are they different and why?

You Are There

This book discusses decomposers in the soil. Imagine you are digging in a forest or a backyard. Write a letter home telling your friends what you have found. What types of insects and fungi could be there? Why are they important? Be sure to add plenty of detail to your notes.

GLOSSARY

adaptation
a change in body or behavior that helps a species survive

carbon
a naturally occurring substance that is present in all living things

cell
a basic building block of all living things

ecosystem
all of the living things in one area

enzyme
a liquid produced by a decomposer to break down waste

leaf litter
fallen leaves on the ground

nitrogen
a naturally occurring substance commonly found as a gas in the air

organic
made of matter that was once living and contains carbon

oxygen
a gas in the air and water that most animals need to breathe in or absorb to survive

photosynthesis
a process where plants use sunlight as energy to turn carbon dioxide and water into food

species
a group of animals that share similar features, lifestyles, and abilities and are able to breed together

ONLINE RESOURCES

To learn more about decomposers and scavengers, visit our free resource websites below.

Visit **abdocorelibrary.com** or scan this QR code for free Common Core resources for teachers and students, including vetted activities, multimedia, and booklinks, for deeper subject comprehension.

Visit **abdobooklinks.com** or scan this QR code for free additional online weblinks for further learning. These links are routinely monitored and updated to provide the most current information available.

LEARN MORE

Hamilton, S. L. *Beetles*. Minneapolis, MN: Abdo Publishing, 2015.

Hirsch, Rebecca E. *Soil*. Minneapolis, MN: Abdo Publishing, 2015.

INDEX

About the Author

Emma Huddleston lives in the Twin Cities with her husband. She enjoys writing educational books, but she likes reading novels even more. When she is not writing or reading, she likes to stay active by running and swing dancing.